O9-BUB-148

WHEELS
AT WORK AND PLAY

ALL ABOUT
SPECIAL ENGINES

For a free color catalog describing Gareth Stevens' list of high-quality children's books, call 1-800-341-3569 (USA) or 1-800-461-9120 (Canada).

Wheels at Work and Play
All about Diggers
All about Motorcycles
All about Race Cars
All about Special Engines
All about Tractors
All about Trucks

Library of Congress Cataloging-in-Publication Data

Stickland, Paul.
   All about special engines / Paul Stickland.
     p. cm. — (Wheels at work and play)
   Summary: Depicts trucks with special functions, such as the moving van, street sweeper, cement mixer, and garbage truck.
   ISBN 0-8368-0425-2
   1. Trucks—Juvenile literature. [1. Trucks.] I. Title. II. Series.
  TL230.15.S749   1990
  629.224—dc20   90-9818

This North American edition first published in 1990 by
Gareth Stevens Children's Books
1555 North RiverCenter Drive, Suite 201
Milwaukee, Wisconsin 53212, USA

First published in the United States in 1988 by Ideals Publishing Corporation with an original text copyright © 1986 by Mathew Price, Ltd. Illustrations copyright © 1986 by Paul Stickland. Additional end matter copyright © 1990 by Gareth Stevens Inc.

Series editor: Tom Barnett
Designer: Laurie Shock

Printed in the United States of America

1 2 3 4 5 6 7 8 9 96 95 94 93 92 91 90

WHEELS
AT WORK AND PLAY

ALL ABOUT
SPECIAL ENGINES

Paul Stickland

Gareth Stevens Children's Books
MILWAUKEE

The furniture is loaded into a van.

The van takes it to a
new home.

The man on the stretcher is ill.

An ambulance takes him to
the hospital.

This truck sweeps the streets.

This cement mixer
pours concrete.

The firefighter pumps water.
It hits the burning house.

The platform goes up
and down.

These people are making a
new road.

They use the rollers to make
it smooth.

The workers empty trash
into trucks.

The person on the platform
fixes the light.

Glossary

ambulance
A truck which takes people to the hospital when they are sick or hurt.

cement mixer
A truck that mixes gravel, water, and cement to make concrete.

concrete
A hard material used to make roads and sidewalks.

firefighter
A person who rides a fire truck and puts out fires.

hospital
A place where people go to get medical help.

stretcher
Something used to carry people who are sick or hurt when they can't walk.

trash
Things people throw out. Garbage.

van
A truck that can be used to carry furniture.

Index

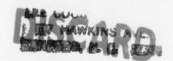